COUNTRY KITCHEN RECIPES

WITH

WHOLE- MEAL FLOUR

MAURICE HANSSEN

First published September 1979
Second Impression November 1979

© THORSONS PUBLISHERS LIMITED 1979

ISBN 0 7225 0565 5

Typeset by Harper Phototypesetters, Northampton.
Printed and bound in Great Britain by
Stanley L. Hunt (Printers) Limited,
Rushden, Northamptonshire.

CONTENTS

INTRODUCTION

Just a few years ago nutritionists believed that there was not much difference between white and wholemeal bread. They were convinced that both were just as good for you, and they considered those who believed otherwise to be harmless but a bit cranky. Indeed, two very famous nutritionists, McCance and Widdowson, in 1956 declared: 'The important thing is that it doesn't seem to matter to your health whether you eat white bread or brown bread — it's only a matter of taste'.

Then three British doctors, Burkitt, Painter and Cleave, discovered that diseases of the gut, appendicitis, diabetes, piles and varicose veins — even cancer — could be linked to the diet of refined foods so prevalent in western civilization. With this, roughage became respectable! Way back in the last decades of the nineteenth century Dr. Allinson said much the same thing when he wrote about the advantages of wholemeal bread, but he was thrown out of the medical profession for such 'ridiculous' ideas. Pioneers often have to pay the penalty for leading the way.

Wholemeal, or wholewheat, flour contains one hundred per cent of the wheat grain, while white flour contains about seventy per cent. The missing thirty per cent comprises not only the bran — which is taken out during the milling process because it colours the flour — but also the wheat germ with its high content of vitamins E and B. There is also a loss of trace elements including the mineral chromium which is thought to be important in the formation of blood.

This is not to say that white bread and white flour are necessarily bad for you — they are good foods — but it is a simple fact of life that wholemeal is better. It is more satisfying and it contains slightly more protein than white flour. And, of greatest importance, wholemeal flour contains essential roughage. So, use it when you can, and if your family are not that enthusiastic, start off mixing some wholemeal flour in with the white flour, gradually increasing the proportion.

All health stores sell several types of stone-ground wholemeal flour, and it is now becoming increasingly available in supermarkets. Try all the types and see which one suits your style of baking the best. Having found it follow the recipes in this book and enjoy eating your way to good health.

SUPER-FAST BRAN BREAD

8 oz (250g) bran
1 oz (25g) dried yeast
1 tablespoonful raw sugar
3 lb (1½ kilos) 100% wholemeal plain flour
1 oz (25g) vegetable margarine or cooking fat
1¾ pt (1 litre / 2 US pints) warm water
1 tablet (100mg) vitamin C
2 teaspoonsful sea salt

Crush the vitamin C tablet and mix with the yeast and the teaspoonful of sugar in a small bowl. Whisk in half a pint of the warm water and leave in a warm place for fifteen minutes when it should froth.

Whilst waiting, put baking tins or baking sheets under the grill to warm.

Put all the dry ingredients into a large bowl. Stir in the rest of the warm water and then the yeast mixture. Mix thoroughly.

Sprinkle some flour over your hands and over a clean

working surface. Tip out the contents of the bowl and knead the mixture for between ten and fifteen minutes. Kneading is a satisfying art in which the dough is pulled and stretched so that the gluten — which is the protein part of the wheat — becomes elastic ensuring that your bread will rise nicely.

HOW TO KNEAD: Spread the dough loosely on the working surface, then fold the far edge of the dough over towards you so that you fold it in half. With the left hand, hold the edge nearest to you, and with the heel of the right hand, firmly push the dough away from you allowing it to stretch at its own speed. Turn the dough one quarter clockwise, fold again in half towards you, and push again. Then carry on, turning, folding and pushing for ten minutes or so.

The dough should now be smooth and resilient to the touch. Divide it into four pieces and, if you want round loaves, mould them into shape with your hands. Put in a warm place and allow to rise for about half-an-hour. If after this time the bread does not look as though it has risen enough, then leave for a little longer.

If you are making loaves in a baking tin, roll out the dough so that it is about three times as long as the tin;

rather like a flat sausage. Fold into three, put in the baking tin, cover loosely with polythene and leave in a warm place to rise.

About fifteen minutes before the rising is due to finish, turn your oven up to the highest setting. When the dough has risen, put in the hot oven and turn the heat down to Gas Mark 7 (425°F/220°C) and bake for about half-an-hour. If the bread looks as if it is browning too rapidly then turn the oven down a bit.

After twenty minutes, open the oven very gently and quickly brush a little oil onto the top of the loaves to give them a professional and glossy appearance.

When ready turn the loaves out onto a cooling grid. You can check the loaves to see if they are properly done by tapping the base with your knuckles, they should sound hollow. Sometimes I find they don't so I quickly put them back in the oven and leave them for another ten to fifteen minutes on a slightly lower heat. Different flours and different ovens react in different ways, but you will soon get used to your own materials.

QUICK WHOLEMEAL BREAD

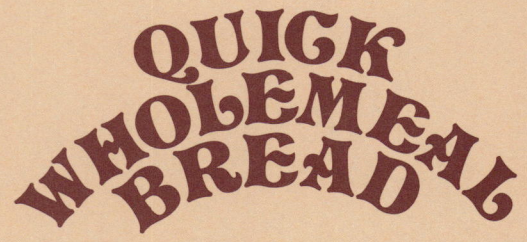

Just follow the SUPER-FAST BRAN BREAD recipe, but without the bran. Use half-a-cupful less water, but everything else stays the same.

Both these breads can be made to look very attractive for special occasions by sprinkling the tops with sesame seeds or cracked wheat or bran before baking.

BANANA BREAD

2 oz (50g) butter
½ cupful raw sugar
1 egg
4 medium-sized bananas
¼ teaspoonful vanilla essence
2 teaspoonsful baking powder
¼ teaspoonful sea salt
½ teaspoonful bicarbonate of soda
¾ cupful chopped nuts (optional)
3 cupsful stoneground wholemeal flour

Cream together the butter and sugar, and beat one egg into the cream.

Mash the bananas with the back of a fork together with two tablespoonsful of water and the vanilla essence. Add this to the butter, sugar and egg mixture and stir in, adding the nuts.

Blend the flour, baking powder, salt and bicarbonate of soda, and gently stir this mixture into the rest. This is a bit sticky, so line the baking tin with greased paper and bake in a moderate oven at Gas Mark 4 (350°F/ 180°C) for between forty and fifty-five minutes.

WHOLEMEAL DATE BREAD

2 eggs
¾ cupful brown sugar
2 teaspoonsful baking powder
3½ cupful wholemeal flour
1 teaspoonful bicarbonate of soda
1 teaspoonful vanilla essence
chopped nuts (optional)
2 cupful chopped stoned dates

Pour two cupsful of boiling water over the chopped dates and then get another bowl. In it, beat the two eggs until they are light and airy. Beating all the time, add to the eggs the sugar, and when creamy, gently sift in the baking powder, one cupful of the flour and the bicarbonate of soda.

Whisk the chopped dates and boiling water together with a fork, add half to the blend, together with the rest of the flour and the vanilla essence.

After stirring this in, pour in the remaining date

mixture together with (if you like) some chopped nuts up to one cupful in quantity. Put the whole mixture into a bread tin and bake in a medium oven at Gas Mark 4 (350°F/180°C) for about an hour.

PUMPERNICKEL

¼ cupful bran
4 cupsful cracked rye (rye meal)
1 cupful cracked wheat
2 teaspoonsful sea salt
2 tablespoonsful vegetable oil
2 tablespoonsful molasses
3 cupsful (approximately) boiling water

Mix all ingredients together into a wet mess. Cover and allow to stand overnight. If the mixture is still sloppy, then add bran little by little until you can shape a nice loaf.

Now roll the loaf in some more bran, wrap in aluminium foil and put on a baking tray with some hot water in a large dish on the shelf below.

Cook gently for about four hours at the lowest temperature you can. For me this is Gas Mark ¼ (225°F/110°C). When done, unwrap, cool completely, rewrap and put in the refrigerator for two days so that the flavours can marry.

(a heavy, juicy bread, full of fibre and flavour)

WHOLEMEAL SCONES

3 cupsful wholemeal flour
2 oz (50g) cooking fat
½ teaspoonful sea salt
4 teaspoonsful baking powder
milk
vegetable oil

Blend the flour with the cooking fat. Add the salt and baking powder and gently blend in enough milk to make a smooth dough (this will need just over half-a-cupful of milk). Roll out to about half an inch thick, cut into rounds with the rim of a tumbler, paint the top with a little vegetable oil and bake in a hot oven at Gas Mark 6 (425°F/220°C) for about fifteen minutes.

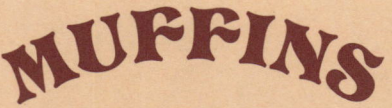

MUFFINS

3 cupsful wholemeal flour
½ teaspoonful sea salt
1 cupful milk
½ oz (15g) baker's yeast
2 oz (50g) vegetable margarine or butter

Half-an-hour before you begin to mix the other ingredients, add a little tepid water to the yeast, cover, and put aside in a warm place so that it begins to bubble.

Bring the milk to the boil and allow to cool to body heat. Blend this with the flour, salt and margarine/butter, and add the mixture to the bubbling yeast, and knead (see first recipe) so that you have a soft dough.

If the mixture is too firm, add a little warm water.

Cover with a cloth and leave in a warm place to rise. This takes about half-an-hour. When risen, knead once more, and then roll out to about half an inch thickness. Cut into circles with the rim of a tumbler or a pastry cutter, bake in a well oiled heavy frying pan or griddle, turning the muffins over so that both sides are well done. Serve piping hot.

WHOLEMEAL BISCUITS

½ cupful molasses
½ cupful vegetable oil
4 cupsful 100% stoneground wholemeal flour
½ cupful milk or skimmed milk
½ cupful raw sugar

Mix all the dry ingredients in a bowl and vigorously beat the other ingredients together in another. Now beat the two mixtures together until they are smoothly blended.

Drop the mixture from a spoon onto an oiled baking sheet so that the biscuits are an inch or two apart. Bake at once in a hot oven at Gas Mark 7 (425°F/220°C) for between ten and fifteen minutes.

This recipe works well as it is, or you can add three teaspoonsful of powdered ginger to make ginger biscuits, or a mixture of one teaspoonful each of cinnamon and ginger with half-a-teaspoonful of allspice to make delicious spicy biscuits.

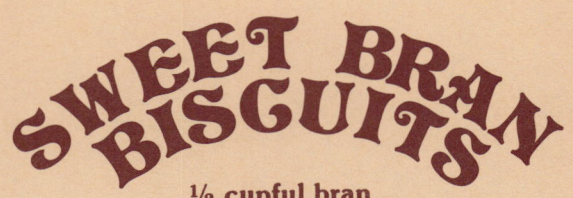

SWEET BRAN BISCUITS

½ cupful bran
½ cupful fine oatmeal
1 tablespoonful raw sugar
½ cupful wholemeal flour
½ teaspoonful ground cinnamon
4 oz (100g) vegetable fat
grated rind of a lemon
water to mix

Blend everything together and make into a firm dough with cold water. Roll the mixture out several times in order to achieve an even consistency of about a quarter inch thick.

Cut into biscuit shapes, prick lightly all over with a fork, put the biscuits on a lightly oiled tin and bake in a hot oven for between fifteen and twenty minutes.

When you remove the biscuits from the oven allow them to become quite cold before handling them because they are very fragile until cool. If you wish to make a variety of biscuits, a handful of seedless raisins or currants can be added to part of the mixture, but take care not to burn the fruit.

HONEY KNOBS

½ cupful raw sugar
½ cupful honey
½ cupful stoned raisins
2½ cupsful bran
⅔ cupful milk
½ cupful finely chopped nuts
2 cupsful wholemeal flour
⅔ cupful butter or vegetable margarine
2 eggs
¾ teaspoonful bicarbonate of soda
½ teaspoonful sea salt
1 teaspoonful baking powder

Beat the eggs well and add them to the melted butter/margarine, sugar and honey, mixing all the time. In another bowl mix together all the dry ingredients. Stir in the milk and then the liquid mixture.

Blend so that everything becomes smooth and then drop onto a well greased baking sheet with a teaspoon, and bake for ten minutes at Gas Mark 4 (350°F/180°C). Try also with a teaspoonful of cinnamon.

BOURGHOL IMFALFAL

½ teaspoonful sea salt
½ lb (225g) crushed wheat
1 large onion (chopped)
3 tablespoonsful vegetable oil
a pinch of saffron

Heat the oil in a thick bottomed pot. Fry the chopped onion until it becomes slippery, then add the crushed wheat and allow both the ingredients to simmer until the onion has lightly browned.

Meanwhile, be soaking the saffron in two cupsful of boiling water. As soon as the onion is ready, add the boiling water with the saffron and salt. Cook, stirring occasionally in a closed pot until the water has been completely taken up by the wheat.

The Arabs use this way of preparing wheat as an alternative to rice, and to give variety. You can serve it with any savoury meal.

WHOLEMEAL PARKIN

1 oz (25g) raw sugar
12 oz (350g) wholemeal flour
8 oz (225g) medium oatmeal
4 oz (110g) butter or vegetable fat
¼ teaspoonful ground ginger
small pinch sea salt
1 teaspoonful baking powder
¼ cupful warm milk
1½ cupful black molasses

Mix together the flour, oatmeal, sugar, ginger and salt. Separately, mix the baking powder in the warm milk, and then stir this into the dry ingredients.

Mix the butter/vegetable fat into the molasses, and mix thoroughly into the rest of the ingredients. Pour out onto an oiled shallow baking tin and bake in a moderate oven for one hour. Then turn off the heat and leave to cool down in the oven for about fifteen minutes before removing.

GRILLED HERRINGS

The herrings you choose should have shiny scales and be firm to the touch.

Clean them thoroughly and wash them. If they happen to have soft roes these are delicious grilled on wholemeal toast by themselves.

Sprinkle the herrings liberally with toasted wholemeal flour and grill under a medium grill (not too close), turning from time to time, until the herrings are well cooked. Then serve with wholemeal bread or toast.

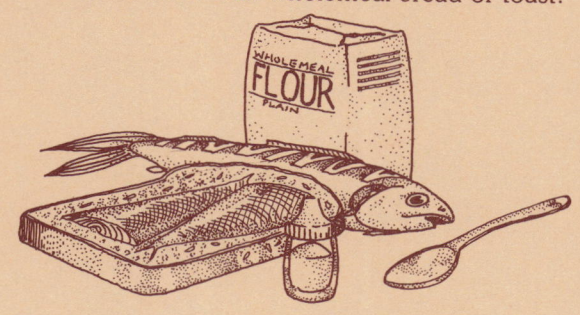

BREAKFAST CEREAL

1 lb (450g) bran
1 lb (450g) stoneground wholemeal flour
3 lb (1¼ kilos) chopped wheat
¾ lb (350g) sultanas
¾ lb (350g) raw sugar
10 fl oz (275g) soyabean oil
6 oz (175g) chopped nuts

Mix everything together thoroughly (if too dry, add a little honey) then put through a mixer. Spread the output from the mixer on a lightly oiled baking tray, and cook in a moderate oven at Gas Mark 3 (325°F/170°C) for half-an-hour. Cool, and then pack into tightly sealed containers.

BREAKFAST CEREAL

Or try:

3 lb (1¼ kilos) 100% wholemeal flour
1 lb (450g) bran
4 lb (2 kilos) chopped cashew nuts
2 lb (1 kilo) runny honey
4 lb (2 kilos) stoned date paste

Mix well together, put through a mixer and cook in a moderate oven for between thirty-five and forty-five minutes.

AUTHOR'S NOTE: On both of the breakfast cereal recipes the manufacturing rights are reserved.

CHAPATIS

½ cupful bran
2½ cupsful fine wholemeal flour
1 teaspoonful sea salt
1 tablespoonful vegetable oil
1 cupful warm water

MAKES TWENTY

Mix two cupsful of the flour and the bran together. Rub in the salt and the oil, and pour on the water, kneading for at least fifteen minutes (see first recipe for how to knead). Make the dough into a ball, cover with polythene and leave for at least an hour (or overnight in the refrigerator).

Lightly flour your working surface, make the dough into about twenty small balls and then roll out, sprinkling more flour as necessary, so that they are round in shape and as thin as pancakes.

Heat up a heavy frying pan or a large, very hot grill. A griddle is ideal. Cook each side for about one minute, pressing down the outside lightly with a fork. Serve at once.

Chapatis can be filled with a savoury filling, in which case you will need a little oil when you cook them.

PURI

Make chapatis (as above) but, instead of cooking them on a griddle, deep fry them in hot vegetable oil so that they swell up and become a beautiful golden brown colour. Serve at once. The result is wonderful!

PARATHA

2 cupsful white flour
2 cupsful wholemeal flour
2 teaspoonsful sea salt
10 tablespoonsful vegetable oil
1½ cupsful water
some oil for cooking

Make the mixture as for the chapatis, and allow to stand. Parathas need more dough than chapatis, so divide it into fifteen parts and form each into a ball. On a floured surface, roll out each ball into a circle, and put two teaspoonsful of oil into the middle of each, and spread it over the surface.

Make a cut from the centre to one edge of each circle, and lift the outside cut edge up with a knife and roll it into a cone. Stand the cone on its base and gently push it into a ball shape again and roll out very gently to about twice the thickness of a pancake.

Cook in a hot oiled frying pan, or on an oiled griddle, adding more oil when the paratha is turned over.

HAMBURGERS

1 teaspoonful sea salt
1 large chopped onion
1 lb (450g) finely minced lean beef
5 tablespoonsful wholemeal breadcrumbs
3 tablespoonsful tomato juice
½ teaspoonful ground black pepper

Mix the ingredients together. Form into hamburger shapes and allow to stand for at least half an hour before cooking in hot oil. Suitably flavoured soaked TVP can be used instead of the beef.

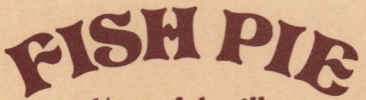

FISH PIE

½ cupful milk
1 teaspoonful sea salt
¼ teaspoonful white pepper
4 slices de-crusted wholemeal bread
4 oz (100g) grated hard cheese
1 lb (450g) cooked white fish
1 large sliced onion

Mash the bread into the milk and allow to stand for ten minutes. Remove all the bones from the fish and break the flesh into flakes (but do not break it up too much). Add the seasonings to the bread and milk mash and then fold in the flaked fish.

Fry the onion rings for a minute or two until they begin to go slippery and then put all the ingredients except the cheese into a heat proof open-topped casserole. Sprinkle the grated cheese over and cook in a moderately hot oven for forty minutes. This can be prepared in advance and cooked when needed.

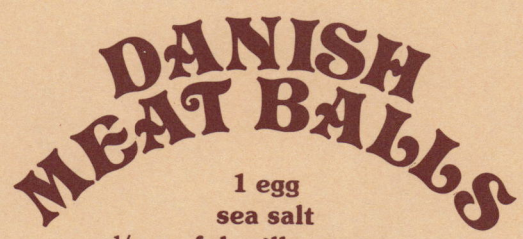

DANISH MEAT BALLS

1 egg
sea salt
½ cupful milk or cream
¾ cupful wholemeal breadcrumbs
3 tablespoonsful grated onion
1 lb (450g) fine minced beef
freshly ground black pepper

Soak the breadcrumbs in the milk. Stir in the egg and onions and finally the meat, seasoning to taste. Mix thoroughly and leave in a cool place for an hour. With wet hands mould the mixture into half golf ball sized shapes. Fry gently in hot oil. Serve hot with gravy or cold on cocktail sticks. Be sure to drain well before serving.

WHOLEMEAL PIZZA

basil
sea salt
bread dough
mozzarella or other cheese
marjoram
tomatoes

A quarter of a pound of wholemeal flour will make an eight inch pizza base. Make it as a flat round disc or yeasted risen dough from one of the bread recipes. The risen dough is traditionally placed on a wooden paddle into a hot, brick oven heated from a wood fire. At home you can make an excellent pizza on a baking tray in a hot oven or by frying the dough in a little oil on both sides and then adding the pizza ingredients and heating through.

You can fill the pizza with a mixture of cooked peeled tomatoes seasoned with basil and marjoram flavoured with any of anchovies, mushrooms, onions, olives, ham and cooked sausage covered with a little cheese.